HOW'S THAT HUMAN?

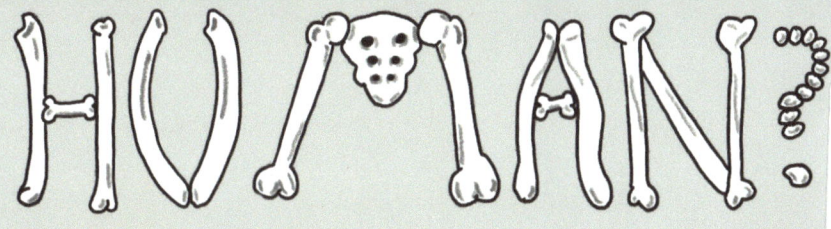

HYDROSPHERE

GEOSPHERE

BIOSPHERE

ATMOSPHERE

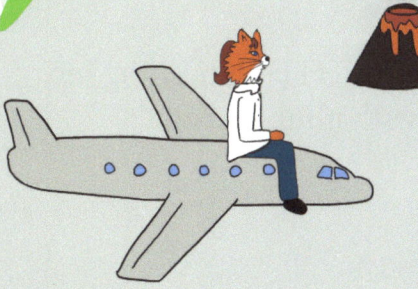

Earth Science

Part 1: Geosphere, Hydrosphere and Atmosphere

Author: Rita Claire

Illustrators: AM Conroy and Kas S.

HOW'S THAT HUMAN?

Cover Design: Rita Claire, AM Conroy and Kas S.
Opening Page: Rita Claire, AM Conroy and Kas S.
Illustrations: AM Conroy and Kas S.

How's That Human Earth Science Part 2 - 1st Edition (softcover)
ISBN 9798853153035

Published by HTH Publishing Inc.
www.howsthathuman.com

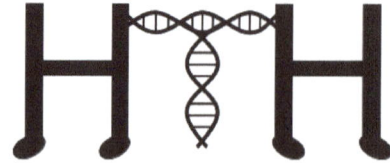

THANK YOU!

<u>Mom and Dad</u>
I want to thank you for always believing in me and supporting my goals in life no matter what they were or where they took me! I love you!

<u>Friends and Family</u>
Thank you for your continued enthusiasm for my ideas! Thank you for not only pushing me but helping me grow, test, and develop this book!

<u>My Team</u>
I couldn't have asked for a better team to make this dream a reality! Your hard work was infectious, teamwork makes the dream work!

<u>Former Students</u>
I can't thank you enough honestly! You pushed me to be better and I am forever grateful for your influence in the teacher I am today!

<u>YOU!</u>
Thank you for taking a chance on a science book that looks at the human side of things and why science is ... well... YOU!

TABLE OF CONTENTS

INTRODUCTION

Humans are intimately linked to the earth and all of the living and non-living parts.
Native Americans have a great appreciation for the earth and all the connections between these parts.

"What are you? From where did you come? I have never seen anything like you." The Creator Raven looked at Man and was ... surprised to find that this strange new being was so much like himself.
- Eskimo creation myth

"Treat the earth well; it was not given to you by your parents, it was loaned to you by your children. We do not inherit the Earth from our Ancestors, we borrow it from our Children."
- Crazy Horse

"The land is the chief; man is its servant."
- Hawaiian Proverb

We return thanks to our mother, the earth, which sustains us. We return thanks to the rivers and streams, which supply us with water. We return thanks to all herbs, which furnish medicines for the cure of our diseases. We return thanks to the moon and stars, which have given to us their light when the sun was gone.
- Iroquois Prayer

"The earth does not belong to man, man belongs to the earth. All things are connected like the blood that unites one family. Man did not weave the web of life, he is merely a strand in it. Whatever he does to the web, he does to himself. The earth is sacred and men and animals are but one part of it."
- Chief Seattle

"What is the best path for humans to follow on the narrow, jagged surface of the earth?"

Mayans answered their own question with, "The balanced, middle path since it avoids excess and imbalance, hence misstepping and slipping, hence misfortune and ill-being."

The Spheres

The earth is a sphere or a round solid figure. We can divide the earth into separate special spheres that describe certain processes or layers that exist on/in/around our planet.

HYDROSPHERE

All the water found on, under and over the surface of the earth.

GEOSPHERE

The earth's insides and crust. This includes landforms, rocks, soils, and plate tectonics.

BIOSPHERE

The region where all living organisms exist.

This includes bacteria, protists, fungi, plants and animals.

ATMOSPHERE

The layers of gases that surround the earth.

It provides protection from space radiation and weather also occurs here.

It is important to understand how all these special spheres interact with one another and how you, as a human, are not only affected by each but more importantly, how you impact them too!

Natural disasters impact human lives every year!

Minerals and resources provide the building blocks for life and enrich human lives.

SPHERES UNITE!

Be on the lookout throughout the book for when the other spheres impact the biosphere!

Functions of Phosphorus

1. Forms bones and teeth
2. Forms DNA and RNA
3. Maintains energy levels
4. Forms cell membranes
5. Keeps nervous system healthy
6. Maintains acid balance
7. Helps oxygen delivery in blood

The Biosphere

Welcome to the final book in the HTH series; Earth Science Part 2 the Biosphere and Environmental Science. Through the other books you have learned that humans are made from the same atoms in star dust (SPACE), that these atoms bond together in many ways to create living and non-living objects (CHEMISTRY), and that forces help shape who we are and what we are capabe of (PHYSICS).

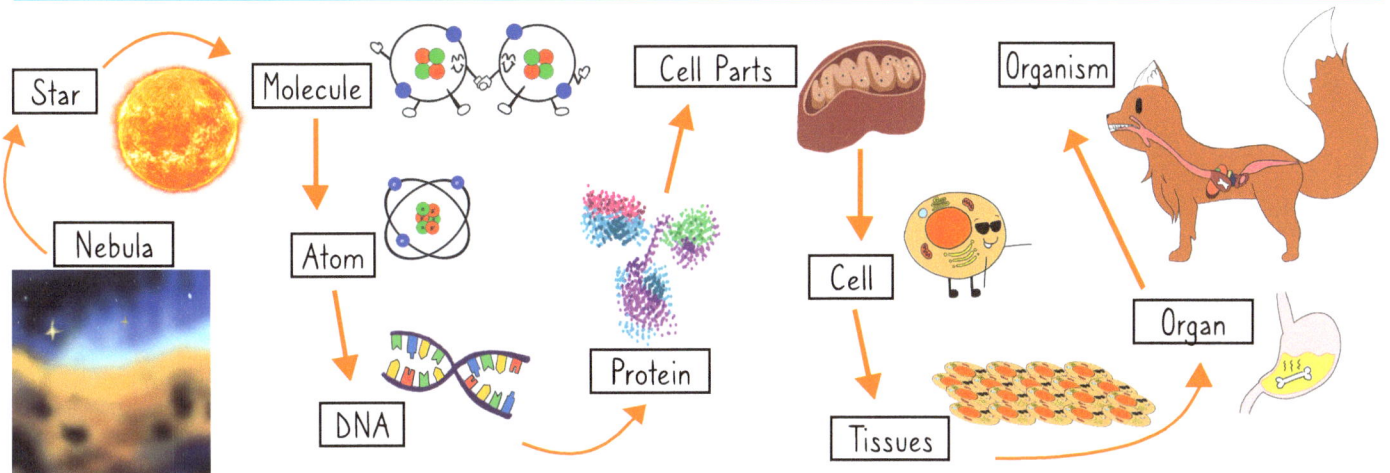

Star → Molecule → Atom → DNA → Protein → Cell Parts → Cell → Tissues → Organ → Organism

Nebula

In this book, you will go on a journey to explore the relationships between living and non-living matter, and these interactions ultimately impact the success of the entire planet!

A single SPECIES of fox exists as a...

POPULATION which lives with other organisms in a ...

COMMUNITY

The community of living organisms and non-living parts exist together as an ECOSYSTEM

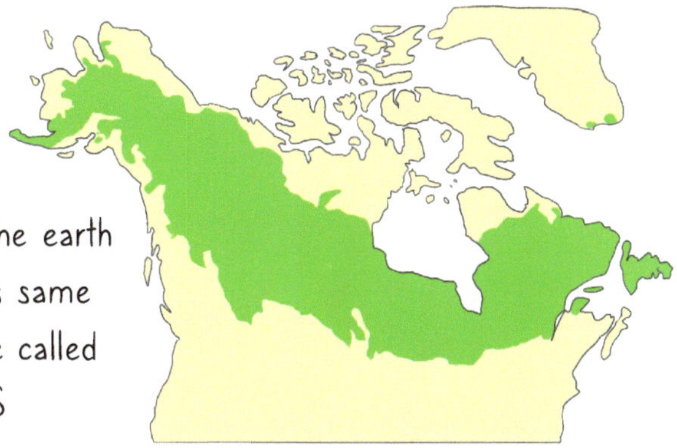

The areas of the earth that have this same ecosystem are called BIOMES

Geological Time Scale

The Geological Time Scale was created to organize the types of organisms that have existed throughout time. Be aware that the following diagram is not to scale. That means the divisions as boxes do not actually represent the amount of time each existed in history as compared to the others.

? QUICK QUESTION:
What is continental drift?

-> The theory that the continents have moved throughout earth's history.

? QUICK QUESTION:
What are mass extinctions?

-> A widespread decrease in the amount and type of living organisms.

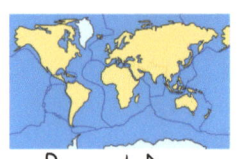
Present Day

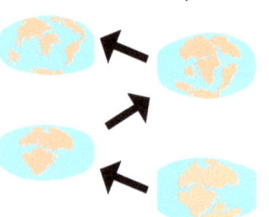

The supercontinent Pangea was formed around 300 million years ago and broke apart into our current continents.

Seas flooded the continents as Rodinia broke up.

Rodinia was a supercontinent that formed around 1 billion years ago!

EON	ERA	PERIOD	
PHANEROZOIC	CENOZOIC	QUARTERNARY	Homo sapiens, 300,000 years ago
		TERTIARY	Earth Cools Down, 2.6 mya
	MESOZOIC	CRETACEOUS	Flowering Plants, 130 mya
		JURASSIC	First Bird Archaeopteryx, 150 mya
		TRIASSIC	Mammals, 210 mya
	PALEOZOIC	PERMIAN	Dinosaurs, 250 mya
		CARBONIFEROUS	Large Forests and Flying Insects, 350 mya
		DEVONIAN	
		SILURIAN	
		ORDOVICIAN	First Land Plant, 470 mya
		CAMBRIAN	Cambrian Explosion, 540 mya
PROTEROZOIC		PRECAMBRIAN	First Multicellular Organisms, 600 mya

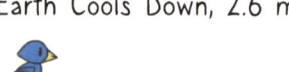

Oxygen Revolution!

Stromatolites, 2500 mya

First Life Forms - Bacteria, 3500 mya

Origin of Earth 4500 million years ago (mya)

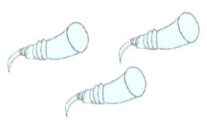

THE PAST: Part 1

QUICK QUESTION: What is an eon?

-> A unit of time a half a BILLION years or more!

The Hadean Eon was the first part of Earth's history when the planet and the moon were still forming, starting about 4.6 billion years ago (that's like 23,000 human generations!) The earth's magnetic field formed and convection currents started working inside the planet causing volcanoes and a very hot surface to form. This eon is named after the Greek god Hades, or the ruler of the underworld.

The oldest rocks found on earth come from this eon. They are called Zircon crystals and they are hidden inside a sandstone conglomerate rock in Western Australia.

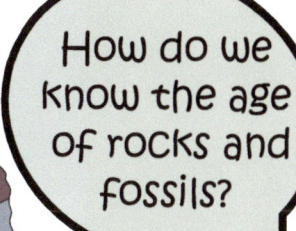

How do we know the age of rocks and fossils?

QUICK QUESTION: What is absolute dating?

-> This type of dating process uses the naturally occurring process of atoms breaking down over time called radioactive decay. This gives a more accurate prediction of age.

Scientist Spotlight

Marie Curie was a Polish born physicist and chemist who is known for her research on radioactivity. She was the first woman to win a Nobel Prize!

QUICK QUESTION: What is relative dating?

-> In general, the oldest layers of rock are on the bottom and the youngest layers of rock are on the top.

Atoms tend to break apart at given rates called <u>half-life</u>, or the time it takes for one half of the starting amount of atoms to break down over a given amount of time.

 5 years  5 more years

Ready to Research?!

What is Carbon-14 dating?
What is the half-life of Carbon 14?

Search Suggestion: what is carbon dating, what is the half-life of carbon-14

Hands-On From Home 🏠

<u>Half-Life Lab</u>
Goal: To simulate radioactive decay.

Materials:
- coins or candy with a print on one side
- bag or shoebox
- data table and graph paper
- writing utensil

Procedures:
1. Place 50 coins or candy in a bag or shoebox and give it a shake (not so hard that you lose any!)
2. Pour the coins/candy out on a table. Time to record some data, you can use the example below on how to set up your data table itself. Pick one side of the coin (heads or tails) or one side of the candy (printed or empty) to be the "DECAYED SPECIES". These are the ones you will remove each time you shake your bag/box. Each time you do shake, it will be considered a "YEAR" of time that has passed. You will record the remaining amount of coin/candy as "UNDECAYED SPECIES".

YEARS	DECAYED SPECIES (Heads)	UNDECAYED SPECIES (Tails)
0	------	50
1	10	40
2	12	28

3. Put your decayed coins/candy off to the side, they are not needed anymore. Place your remaining undecayed coins/candy back in your bag/box, give it a shake and pour it out again. Record your decayed and undecayed species numbers. Remove your newly decayed coins/candy and only place your remaining undecayed coins/candy back in your bag/box.
4. Repeat Step 3 until you have no remaining undecayed species.
5 Graph your "UNDECAYED SPECIES" numbers over time. On the x-axis place your "YEAR" values, or all the shakes it took. The y-axis will be the number of undecayed coins/candy you had over the years. Here is an example graph:

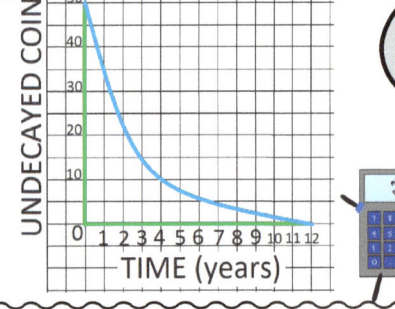

This is called an exponential curve!

The Archean Eon

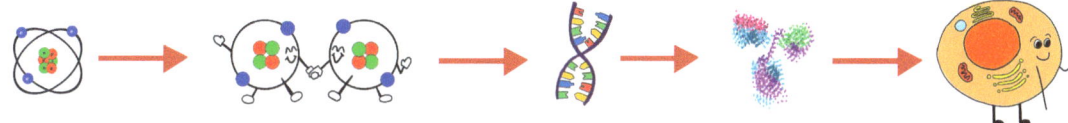

The Archean Eon lasted from 4000 to 2500 million years ago. During this long period of time, the Earth's crust cooled, continents and oceans formed, and life first appeared. This eon is named after the Greek word for beginning.

In order for life to start, a couple of things happened first. Cells, or what we consider the basic building blocks of life, are made out of atoms that bond together in specific ways. These atoms share or give and take electrons with each other depending on what type of elements are involved. When elements in the Archean started forming bonds, they were able to form the major molecules of life, or what we call biomolecules. These biomolecules found ways to interact with each other, including the passing of information, and eventually the first cell was formed.

Want to learn more? Find information in the How's That Human? Chemistry Book!

Three major biomolecules are involved in the central dogma (explanation) of life, which is a theory that genetic information is passed from DNA to RNA to Protein.

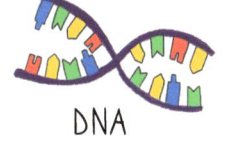 DNA   RNA  Protein

The code for every part of your body is contained in the order of just 4 possible nucleotides that bond together to create a strand of DNA. Each nucleotide contains a phosphate, a sugar, and a base.

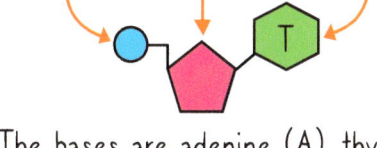

The bases are adenine (A), thymine (T), guanine (G) and cytosine (C). These bases have specific partners they bond with to create the ladder steps within DNA.

A pairs with T
C pairs with G

The code in DNA must be turned into a protein, but in order to do that it must first be turned into RNA through a process called <u>transcription</u>. This nucleic acid is allowed to travel out of the nucleus and to the ribosomes where proteins are made. You will learn about the types of cells and cell parts next!

The code is transcribed into RNA with one difference, RNA has the base U (or uracil) in place of T and it still pairs with A.

The DNA code or message carried by RNA is turned into a protein through a process called <u>translation</u>. Basically the code is translated into a different language, like English into Spanish.

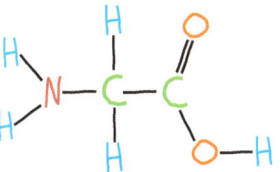

The amino acid Glycine matches the bases: GGG, GGC or GGA.

Proteins are made from molecules called amino acids, and each amino acid matches up with three specific bases in the DNA (now RNA).

The First Cell

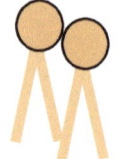

A fourth biomolecule, called a lipid, is a critical component when building a cell. They form an outer membrane or barrier to hold things inside the cell. They are great at keeping things out of the cell as well, kind of like a moat around a castle! Once lipids formed around the other biomolecules, a cell was created. As life grew more and more complex, other structures formed inside the cells to complete different jobs so the cell could survive. The first cell to form was simple, and some of these organisms still exist today!

QUICK QUESTION:
What is a prokaryotic cell?

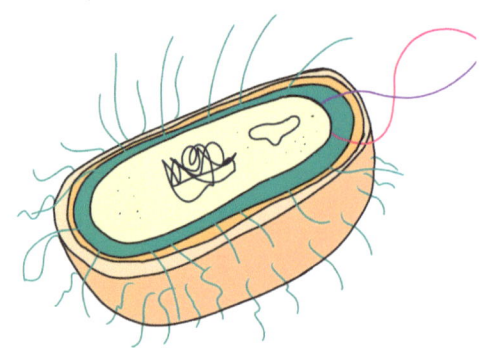

-> A simple cell with no nucleus or specialized cell parts called organelles. These organisms make up two kingdoms of life.

Introducing the first Kingdom of Life, Archaebacteria!

These bacteria are prokaryotes that like living in extreme environments and do not need oxygen. Oxygen levels at the beginning of Earth's history were very low; the atmosphere was primarily carbon dioxide. The environment was hot and harsh, so these organisms really do represent the oldest life on earth!

Methanogens - make methane, live in smelly swamps

Halophiles - salt lovers, thrive in salty conditions

Thermophiles- like it hot, live in extreme temperatures/conditions

Introducing the second Kingdom of Life, Eubacteria!

These are also prokaryotic cells and are the most commonly found organisms on earth. They have specific cell shapes that are used to classify them. Several of these exist on your skin and inside your body!!

Round "Coccus"

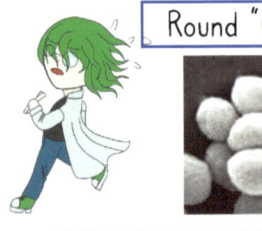

IMPORTANCE
- maintain health of organisms
- vital parts of ecosystems
- human diseases

Rod-like "Bacillus"

 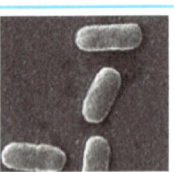

Comma Shaped "Vibrio"

 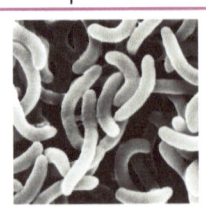

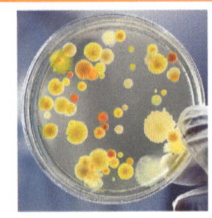

Spiral "Spirilla or Spirochete"

 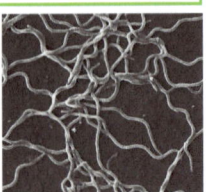

Fancy Cells

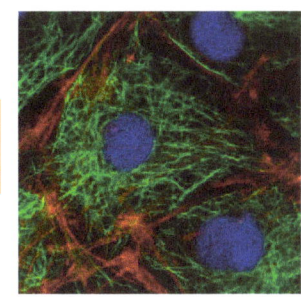

QUICK QUESTION: What are eukaryotic cells?

-> Cells with special parts called organelles. Organelles have their own membranes and specific jobs within the cell.

The second type of cell to appear was the eukaryotic cell. This is the cell found in all the rest of the Kingdoms of Life.

Animal Cell

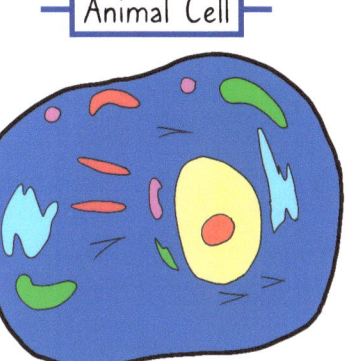

Plant cells have three distinct difference from animal cells:

a cell wall

chloroplasts

a large vacuole

Plant Cell

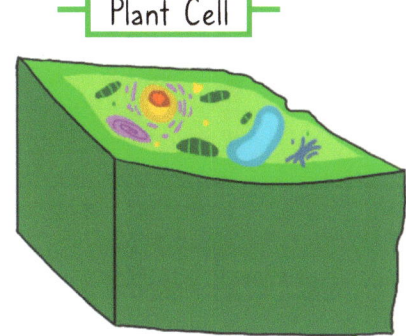

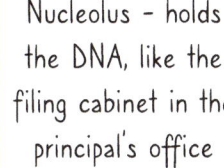

Ribosomes - convert the message in DNA, carried by the RNA, into proteins, like the lunch ladies in the cafeteria

Cell Wall - provides support and protection for the cell, like the fence around a school yard

Nucleus - boss of the cell, like the principal of a school

Nucleolus - holds the DNA, like the filing cabinet in the principal's office

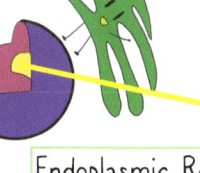

Vacuole - a storage for the cell, especially water in plant cells, like the closets in the school

Endoplasmic Reticulum - holds the ribosomes that make the proteins, and makes sugars and lipids, like the lunchroom at school

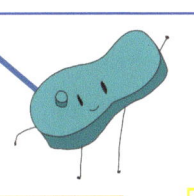

Lysosome - cleans the cell by digesting and recycling proteins and toxins, like the janitor of a school

Mitochondria and Chloroplasts - provide the energy for cells, like the boiler room in a school

Cytoskeleton - microfilaments and tubules that provide structural support for the cell, like the walls of a school

Design your own cell. Don't forget to share with the HTH community via email or social media. Email any project to submissions@ howsthathuman.com or use these hashtags #howsthathuman #hthscience

CREATIVE CORNER

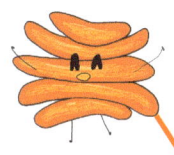

Golgi Apparatus - packages and transports proteins and lipids throughout the cell, like the post office with delivery trucks bringing supplies to school

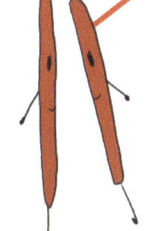

Cell Membrane

The cell membrane is a very important part of any living organism.
If we take a closer look, we can see all kinds of biomolecules working together!

Receptor proteins receive chemical messengers from other cells and organs like hormones.

Glycoproteins are proteins that have a chain of sugar molecules (carbohydrates) hanging off like a necklace to identify the cell.

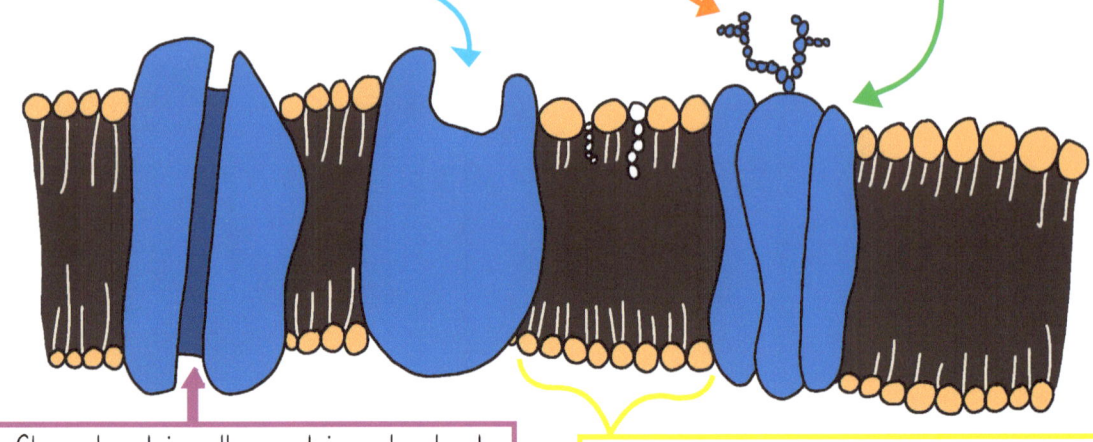

Channel proteins allow certain molecules to pass through that the lipids won't allow.

Phospholipids serve as a barrier, only certain things can pass through like water.

TYPES OF MEMBRANE MOVEMENT

PASSIVE
No energy is needed for molecules to move across the membrane. Molecules naturally move from areas where they are tightly packed or high concentration to areas where there is more room per molecule or low concentration.

ACTIVE TRANSPORT
Sometimes your body needs more of a certain type of molecule and we have to force them to become more concentrated. Not only is a membrane protein needed, but so is energy! Your body has to burn stored energy from food to perform this task.

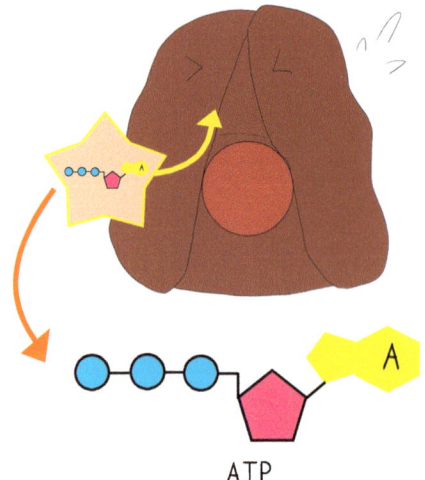

DIFFUSION
The natural movement of small molecules across a membrane to spread out.

FACILITATED DIFFUSION
A membrane protein is used to move bigger molecules.

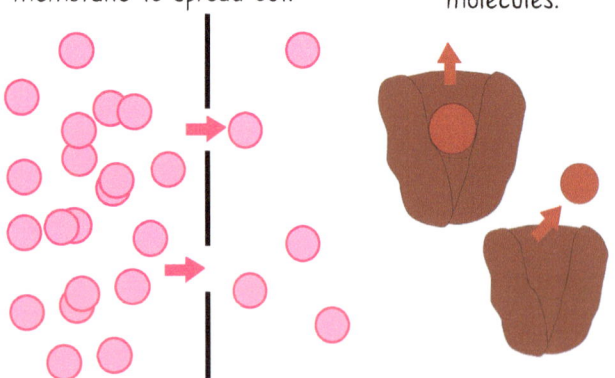

ATP
Adenosine triphosphate is a special molecule in your body that stores the energy you get from breaking down food!

10

Water is a molecule that can freely move across the cell membrane, and this motion is given its very own name!

Water wants to create a balance between itself and the other types of molecules on both sides of the membrane. It will move so the concentrations are equal. However, the cell may find itself in some tricky situations:

QUICK QUESTION:
What is osmosis?

-> The movement of water molecules across a membrane from an area with too many water molecules (high concentration) to an area with not enough water molecules (low concentration).

Hypotonic
The cell is placed in pure water. The water rushes into the cell to try and equal out but the cell bursts.

Isotonic
The cell is in a solution with the same amount of water and other molecules on both sides of the membrane and stays the same size.

Hypertonic
The cell is placed in a solution with low water, the water inside the cell rushes out to try and equal out but the cell shrinks.

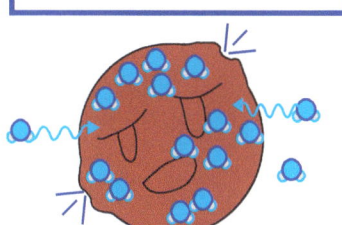

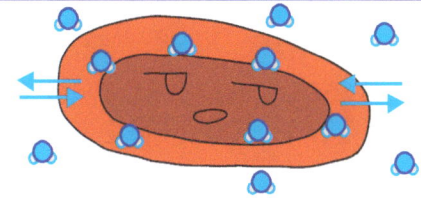

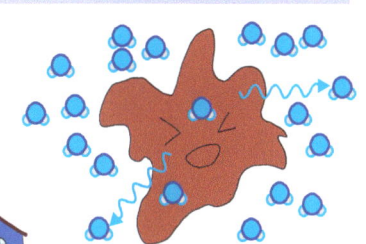

Hands-On From Home

Egg Osmosis Lab
Goal: To model movement of water across a cell membrane using an egg.

Materials:
- 2 raw eggs
- vinegar, water, and corn syrup
- 2 cups
- writing utensil, journal, picture-taking device

Procedures:
Day 1 - Place each egg into a cup and cover with vinegar. Leave overnight for the vinegar to react with the eggshell breaking it down.
Day 2 - Carefully wash each egg, lightly scrubbing any leftover shell off the membrane of the egg. Observe the shape of your egg (take pictures). Clean your cups and fill one halfway with water and the other with corn syrup. Place an egg into each cup (make sure they are completely covered) and let sit overnight or up to 2 days.
Day 3/4 - Observe the shape of your eggs in each cup, take pictures and write down any observations you may have. Which egg was in a hypertonic solution and why? Which egg was in a hypotonic solution and why?

Ready to Research?!
What happens to your fingers after a long time in water, like swimming or a bath?

Search Suggestion: fingers wrinkle bath, fingers in water

DID YOU KNOW:
Salmon have a special protein in their gills that allow them to go from salt water to fresh water to lay eggs?!

Cell Life Cycle

Now that a cell was successfully established with special molecules designed to carry information, the next step was to figure out how to pass on that information and create more cells. Cells were and are able to form, mature, and age just like you. They also figured out a way to reproduce themselves as part of what we call the cell life cycle.

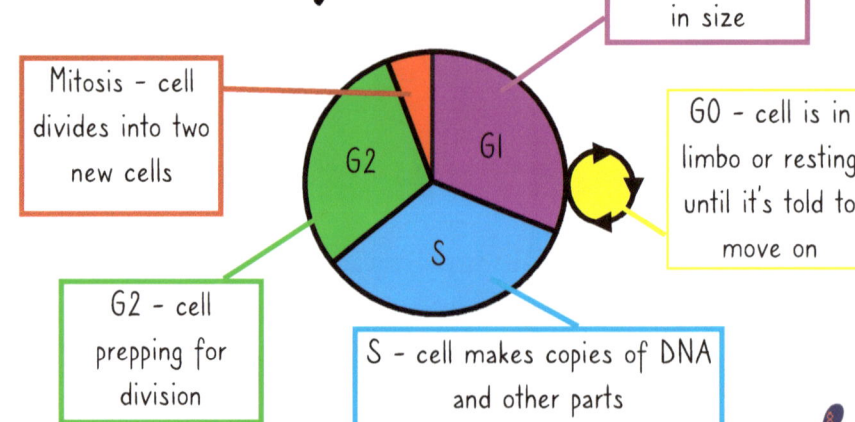

Mitosis - cell divides into two new cells

G1 - cell grows in size

G0 - cell is in limbo or resting until it's told to move on

G2 - cell prepping for division

S - cell makes copies of DNA and other parts

QUICK QUESTION:
What is a chromosome?

-> DNA wound around proteins in tight coils so it can be easily separated into the two new cells. Each chromosome has two copies of the DNA.

The cell goes through a lot of preparation in the G1, S, and G2 phases of its life cycle and together we call these phases <u>interphase</u>. Once the cell has made two copies of the DNA and doubled up on all other cell parts, it can divide quickly into two identical daughter cells. The process of cell division is called <u>mitosis</u>, and we can actually see it happen by watching the chromosomes and their wild dance!

| Prophase (P) the DNA winds up into chromosomes | Metaphase (M) the chromosomes line up in the middle of the cell | Anaphase (A) the chromosomes are split in half and pulled to the opposite sides of the cell | Telophase (T) the cell starts to pinch off into two cells |

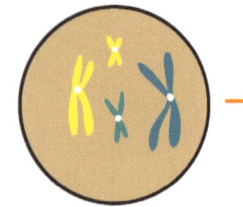 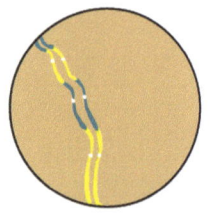 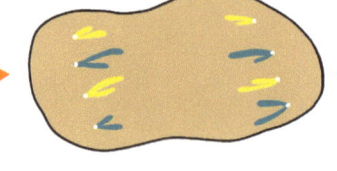 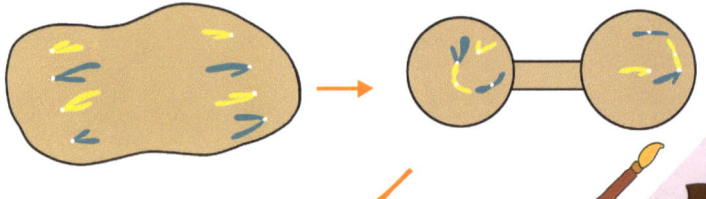

Cytokinesis complete separation into two identical daughter cells

Data Dive

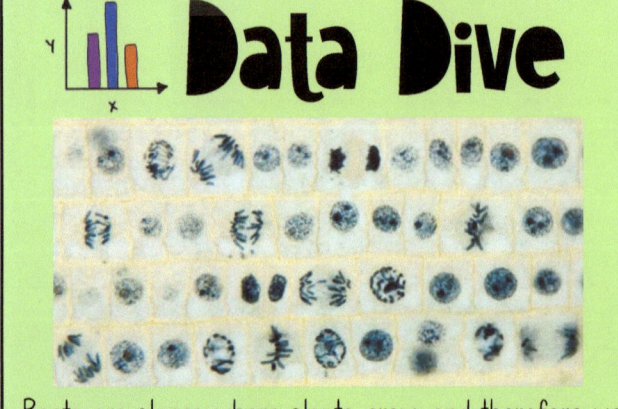

Roots are places where plants grow, and therefore we can catch the stages of mitosis. Pictured are onion root tip cells with the DNA dyed with blue for us to see it. Can you find cells of each stage of mitosis?

CREATIVE CORNER

Create your own Mitosis Flipbook! Show the chromosome dance through PMAT using 2 sets of chromosomes that are different colors. Share a video of your flipbook to be shown on the HTH website! submissions@howsthathuman.com

The Past: Part 2

During the next 2000 million years, the types of different life on Earth exploded. The supercontinent Rodinia broke up into continents that were surrounded by oceans. In the Proterozoic eon, life made a huge leap forward and cells started to work together and form multicellular life forms. Luckily for us, fossils start to appear during this time in rocks.

QUICK QUESTION:
What are multicellular organisms?

-> Organisms made of many cells that take on specialized jobs.

QUICK QUESTION:
What are fossils?

-> The preserved remains or traces of ancient organisms.

MOLD - impression or imprint

TYPES OF FOSSILS

CAST - a mold filled with sediment (3D)

HOW DO FOSSILS FORM?
1. Buried in sediments and turned into rock with heat and pressure. They can completely dissolve and only leave an imprint or impression in the stone.
2. Traces of activity can be buried in sediments and preserved like footprints.
3. Encased in ice or amber or tar pits.

PETRIFIED - tissue replaced with minerals

 **Data Dive**

What type of fossil is pictured?

TRACE - evidence of activity

CARBON FILMS - thin layer of carbon left behind

PRESERVED - in ice, amber, or tar

Energy

With new types of life comes new ways of getting the energy to be able to accomplish everyday tasks. In the beginning, organisms used the minerals in their environment to get energy. They would release energy by performing chemical reactions, something that we still see today at hydrothermal vents and hot springs.

At one point, some lucky organism figured out how to use the energy directly from the sun, and new chemical reactions were created in a process we call photosynthesis. Chloroplasts absorb energy from the sun's electromagnetic waves and rearrange the atoms of water and carbon dioxide.

Want to learn more? Find information in the How's That Human? Chemistry, Physics and Space Books!

The energy released during this process is stored in the new bonds of the compound glucose, a very important carbohydrate. Carbohydrates are yet another important biomolecule!

Oxygen for these organisms is actually waste, but this waste forever changed the world as we know it! Oxygen built up in the atmosphere and new organisms appeared that could use it for their own energy process. These organisms break down glucose with oxygen to release the stored energy in a set of chemical reactions called cellular respiration. This is what humans and animals do!

This leads us to our next Kingdom of Life, Protista. This kingdom has a wide variety of single-celled and multicellular life forms that are all eukaryotic cells (cells with all the organelles). The groups within the kingdom are divided based on how they get their energy.

ANIMAL-LIKE
These protists have to eat something and break it down to release energy, just like humans. They are called protozoa or "first animal".

PLANT-LIKE
These protists are called algae and get energy with photosynthesis.

FUNGUS-LIKE
These protists are organisms that break down dead organisms and are called molds. Slime molds have been shown to act like road engineers!

AMOEBA **FLAGELLATES**

 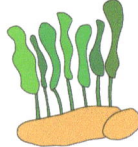 

RED, GREEN OR BROWN ALGAE

WATER MOLD

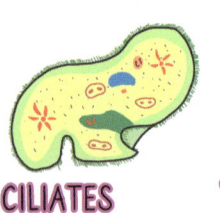 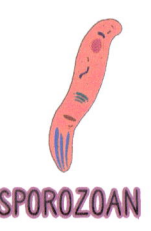

CILIATES **SPOROZOAN**

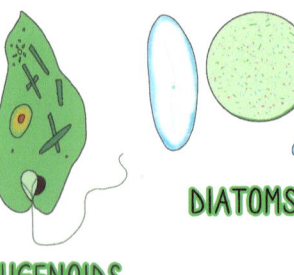 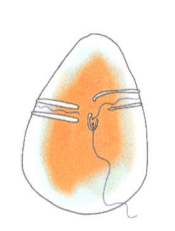

EUGENOIDS **DIATOMS** **DINOFLAGELLATES**

SLIME MOLD

As organisms moved from oceans to land, yet another Kingdom of Life evolved.
As organisms die, their biomolecules are recycled back into the soil by the Kingdom Fungi.
These organisms are eukaryotic and mostly multicellular, save for one special organism – yeast!

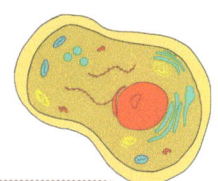

This kingdom is divided up into groups based on the shape of their reproductive structure:

ZYGOMYCOTA
These fungi are very small, grow quickly, and are commonly called bread molds. Their reproductive parts are mostly microscopic.

ASCOMYCOTA
This group are called sac fungi because of the cup shaped structures.

BASIDIOMYCOTA
These are the stereotypical mushroom structures we all think of or eat. The club like structure has many different forms.

Hands-On From Home
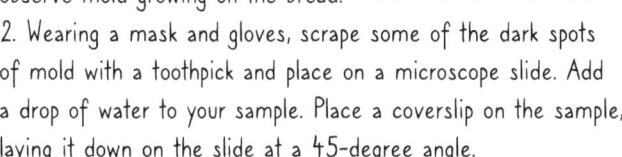

Bread Mold Microscope Lab
Goal: To view a fungal reproductive structure.

Materials:
- microscope, slides, coverslips, droppers, toothpicks
- slice of bread, sealable plastic bag, water
- safety gear = gloves and a mask
- writing utensil, journal

Procedures:
1. Sprinkle the slice of bread with water and place it in a sealed plastic bag. Leave the bag in an undisturbed place for one week or until you observe mold growing on the bread.

2. Wearing a mask and gloves, scrape some of the dark spots of mold with a toothpick and place on a microscope slide. Add a drop of water to your sample. Place a coverslip on the sample, laying it down on the slide at a 45-degree angle.
3. Observe the reproductive structures under a microscope and record your findings. Wash your slide (and your hands after!)

Pond Water Microscope Lab
Goal: To observe freshwater protists.

Materials:
- sample(s) of pond water
- microscope, slides, coverslips, droppers, toothpicks
- safety gear = gloves
- writing utensil, journal, picture-taking device

Procedures:
1. When looking for a sample, find water with algae, or green gooey stuff. Wear gloves just in case.
2. Using your dropper, try and suck up a small chunk or string of green stuff with pond water and put it all on a slide. Place a coverslip on the sample, laying it down on the slide at a 45-degree angle.
3. Observe your sample. Look for moving animal-like protists and the amazing variety of plant-like protists. Web serach each group of protists to see images and names of the different types!

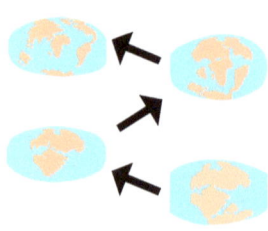

The Past: Part 3

We currently live in the Phanerozoic Eon, which has seen the most growth in what we call biodiversity, or the different types of life. The last 550 million years of Earth's history to the present have witnessed the appearance of so many unique organisms.

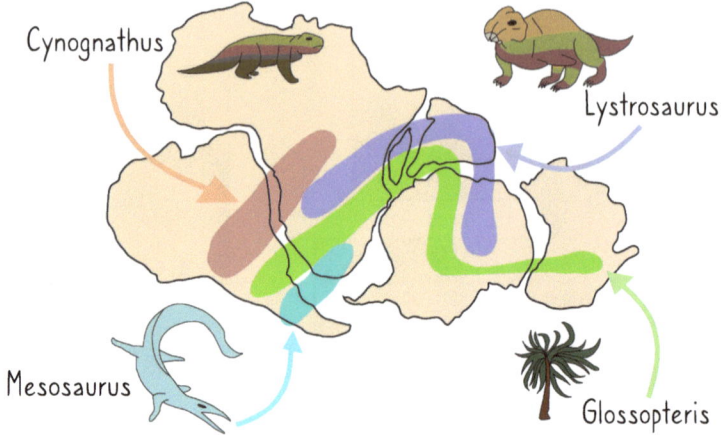

Cynognathus

Lystrosaurus

Mesosaurus

Glossopteris

As the earth was changing over time, unfortunate events called mass extinctions happened at several points. We are discovering more and more about these events and they are not only due to meteor impacts and geological catastrophes like supervolcanoes, but also climate change.

During this time we have seen the continents drift and form the single land mass of Pangea, and then split apart and move into the current positions of our continents today. We have found evidence of Pangea across the continents in many ways, showing how they fit together like puzzle pieces. There are animal and plant fossils, as well as the same rock types found across the continents.
Can you see the continents?

Ready to Research?!

Are we currently living in another mass extinction?

Search Suggestion: mass extinctions, current mass extinction

DID YOU KNOW:
The Permian extinction known as the "Great Dying" killed 95% of ocean life and 70% of land organisms?!

Scientist Spotlight

Gerta Keller is a geologist and paleontologist who was born in Liechtenstein and raised in Switzerland. She has won awards for her research on mass extinctions related to volcanoes and climate change, challenging the meteor hypothesis.

Adaptations

As life moved from the oceans to land and came back from mass extinctions, it turns out successful organisms had adaptations.

QUICK QUESTION:

What is an adaptation?

-> A physical feature or behavior that allows an organism to survive better than others of its kind.

DID YOU KNOW:

There are frogs in Canada that basically cryofreeze themselves over the winter and come back to life in the spring?!

PHYSICAL

These adaptations are structural parts of the organisms like body parts or color.

Ready to Research?!

What adaptations were needed for survival on land?

Search Suggestion: adaptations, adaptations for land

Who wore it better? Vote in the online poll on the HTH website for best camouflage!

Hands-On From Home

BEHAVIORAL

These adaptations are an action or instinct that an organism performs.

PHYSIOLOGICAL

These adaptations are using chemicals or chemical reactions from within the organism.

Thumb Adaptation Lab
Goal: To show how an adaptation increases your chance for survival

Materials:
- pieces of paper
- scissors
- shoes with laces
- masking or medical tape
- timer

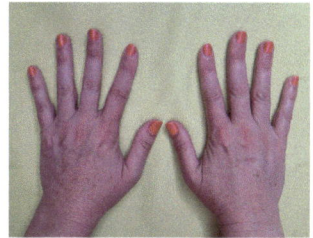
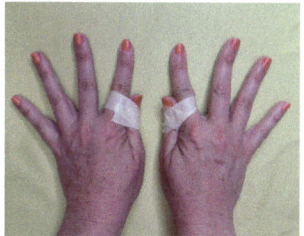

Procedures:
1. Time yourself completing the following tasks normally:
    a. folding a paper airplane
    b. cutting a paper heart
    c. untying and tying a shoelace
2. Tape your thumbs to your pointer finger on both hands and repeat the tasks in Step 1. Compare your times with and without thumbs.
3. What is the advantage to having an opposable thumb?

Eras and Periods

Life appeared so quickly during the Phanerozoic Eon that we had to divide the eon into more specific time frames called eras. Each era of the Phanerozoic had to be further divided into smaller time chunks called periods.

Paleozoic "Ancient Life" Era

EON	ERA	PERIOD
PHANEROZOIC	CENOZOIC	QUARTERNARY
		TERTIARY
	MESOZOIC	CRETACEOUS
		JURASSIC
		TRIASSIC 💀
	PALEOZOIC	PERMIAN
		CARBONIFEROUS
		DEVONIAN
		SILURIAN
		ORDOVICIAN
		CAMBRIAN
		PRECAMBRIAN

- there was a large mass extinction, Pangea formed
- huge forests take over the land, flying insects and reptiles appear in the fossil record, coal starts to form
- flightless insects, spiders, and amphibians are successful
- jawed fish and land plants take off
- the first vertebrate appeared
- there was an explosion of life, trilobites and mollusks

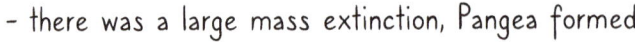

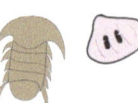

The first era was still apart of the Proterozoic Eon called the Precambrian.

Conditions
- cold
- glaciers
- called "snowball earth"
- oxygen in the atmosphere

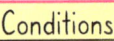

Life Forms
- sponges
- jellyfish
- segmented worms
- protozoa

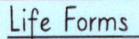

The Final Kingdoms

Welcome to the Kingdom Plantae! The first plants to appear reproduced with spores, but eventually seeds evolved. With flowers, plants could spread their seeds in many ways.

No Flowers

Flowers – plants that reproduce with flowers are divided into two groups based on their body parts.

Plants that reproduce with spores are ferns and mosses.

Plants that reproduce with seeds in cones are gymnosperms.

Monocots have one food storage in the seed and parallel veins in leaves.

Dicots have two food storages in the seed and netted leaf veins.

Welcome to the Kingdom Animalia! This kingdom has a huge amount of different organisms that we have to classify based upon characteristics that they do or do not share.

QUICK QUESTION:
What is an invertebrate?

-> An animal with no backbone, approximately 90% of all life on earth!

?

QUICK QUESTION:
What is a vertebrate?

-> An animal with a backbone, only 10% of all species on earth!

Sponges (Porifera)

Jellyfish (Cnidarians)

Cold-blooded

Fish   Amphibians   Reptiles

Worms (Annelids)

Starfish (Echinoderms)

Warm-blooded

Birds are divided into 23 groups based upon several characteristics, like the ability to fly.

Mollusks – this group all have a muscular "foot" they use to move around, how is it different between them?

Mammals are divided into groups based on how they reproduce. Placentals are like humans, marsupials raise young in a pouch, and monotremes lay eggs!

Bivalves   Cephalopods   Gastropods

Arthropods – this group all have jointed legs, but how many?

Crustaceans   Centipedes   Spiders   Insects

Placentals   Marsupials   Monotreme

Mesozoic and Cenozoic

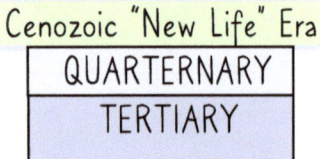

Mesozoic "Middle Life" Era
CRETACEOUS ☠
JURASSIC
TRIASSIC

This era is known as the Age of Reptiles. Pangea broke up and the mass extinction that killed the dinosaurs ended the era.

- flowering plants take over

- dinosaurs dominate, birds and conifer trees are first seen

- reptiles take over and the first mammals appear

Cenozoic "New Life" Era
QUARTERNARY
TERTIARY

In this era the Earth evolved into its final look of today, even still moving. The tectonic plate of the western U.S., called the Pacific Plate, is moving northwest at a rate of 3-4 inches or 7-11 centimeters a year!!

With large reptiles gone, mammals began to dominate, the earth cooled during an ice age, and grasses became a common plant and important food course.

These mountain ranges were built: Alps, Carpathians, Himalayans, Atlas and Rocky Mountains.

Humans appeared with many other upright walking beings called hominids. Humans began to hunt, harness fire, spread and migrate, and eventually learned to stay put.

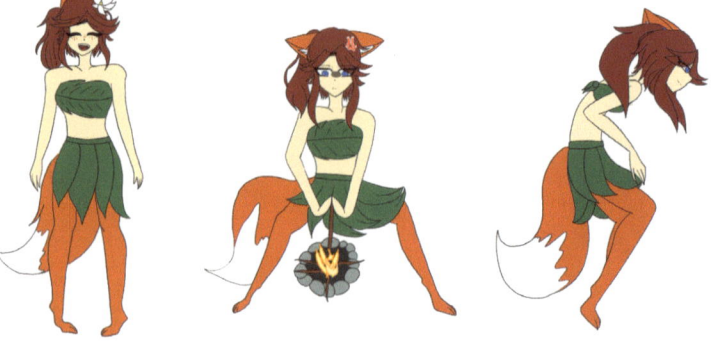

They domesticated animals, formed agricultural practices, began trading with each other, and invented culture through music, art, and stories forever changing the earth yet again...

The Present

Welcome to the present! Hope you enjoyed your exploration through the past. As you have seen so far, there are many characteristics of life. In general all living organisms:

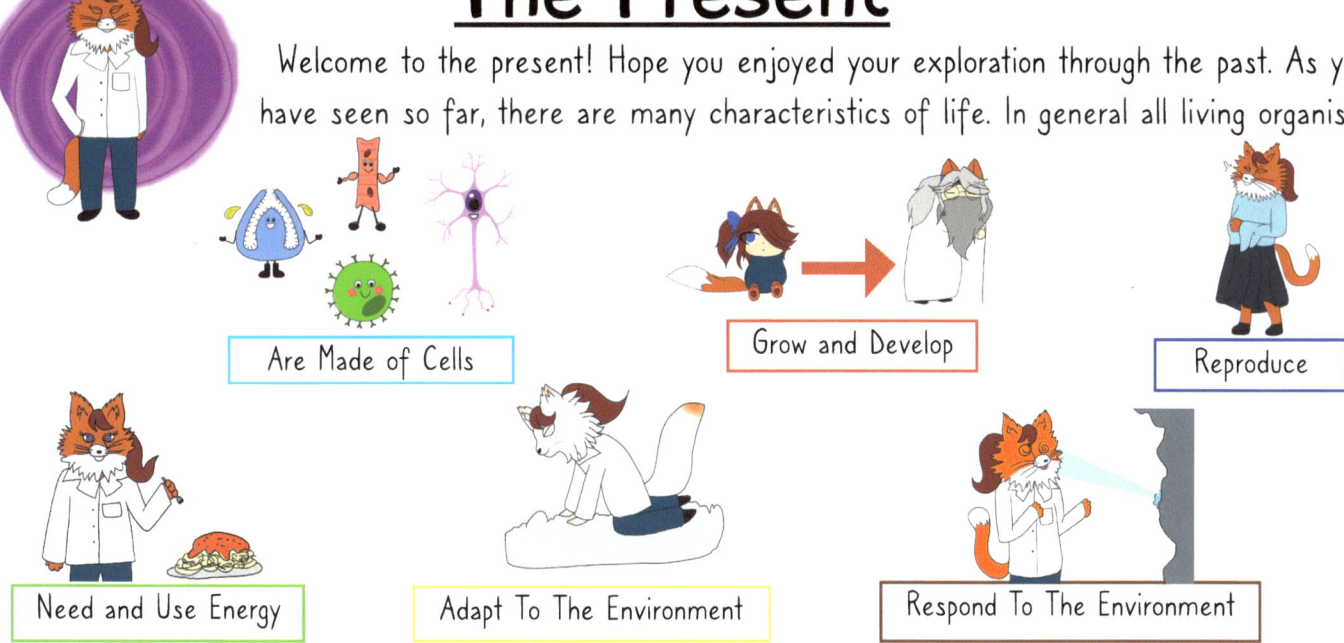

Are Made of Cells

Grow and Develop

Reproduce

Need and Use Energy

Adapt To The Environment

Respond To The Environment

Ultimately survival comes down to the body maintaining a balance of all body systems in the organism.

QUICK QUESTION: What is homeostasis?

-> A state of balance for all body systems. This is needed for life to survive.

Let's look at how humans accomplish the characteristics of life!
Homeostasis is achieved by all the biological systems in your body working together...

Muscular System

Moves (1) the bones at joints with the help of tendons and ligaments, (2) the blood, other bodily fluids, and gases, and (3) food through the digestive system. They protect, support, and generate heat.

Skeletal System

Supports and protects the body and organs, allows movement, gives the body its shape, stores minerals, and makes red and white blood cells.

Integumentary (Skin) System

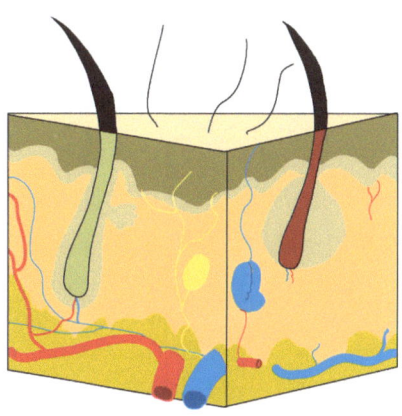

The many parts of the skin contribute to the insulation and protection of the body, temperature and water regulation (sweating), synthesis of Vitamin D, and detection of stimuli in the environment (touch, pain).

CREATIVE CORNER

Are viruses alive? Join the discussion on the HTH website!
www.howsthathuman.com

The Human Body Continued

Nervous System

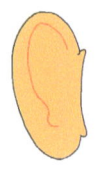 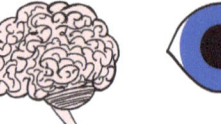 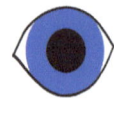

Controls and regulates all body systems. The brain is divided into specific regions in charge of specific jobs. The nerves deliver the communication to the rest of the body. Specialized organs sense the environment for you.

Endocrine System

Works with the nervous system to regulate the body, communicates using hormones (messengers), and keeps the body in balance.

Digestive System

Breaks down nutrients chemically and physically, absorbs nutrients, and eliminates wastes.

Reproductive System

Allows for making offspring so life can continue. Communication from the endocrine system helps develop a baby through pregnancy and birth.

Urinary System

Gets rid of wastes and excess water.

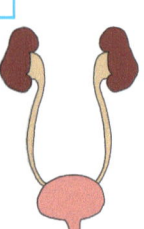

Immune System

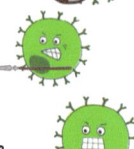

Cells, tissues, and organs working together to fight infection and disease.

Hands-On From Home

Human Body Lab

Goal: To test your body's homeostasis through exercise.

Materials:
- space to exercise
- timer
- writing utensil, journal

Procedures:

1. Pick 2 different exercises to test your breathing rate and heart rate. Hypothesize which one will increase the rates the most.
2. Take your "RESTING" heart rate by counting the beats of your heart for 6 seconds, then multiply that number by 10 to get your BPM or beats per minute. Do the same for your breathing rate. Record these in a data table like the one below. (example data)
3. Perform your first exercise for 30 seconds. Immediately record your heart rate and breathing rates after in the data table.
4. Rest for a few minutes to let your body get balanced again.
5. Repeat #2 and #3 for the second exercise.
6. Compare the data! Was your hypothesis supported or not?

	Heart BPM Before	Heart BPM After	Breathing Before	Breathing After
Exercise 1 Jogging	80	150	20	50
Exercise 2 Sit-ups	80	130	20	40

Lymphatic System

Follows the tubes of the circulatory system collecting excess fluid to recycle. Participates in protecting the body by transporting and making immune system cells.

Circulatory "Cardiovascular" System

Transports nutrients to and takes wastes away from all systems of the body. The heart is the pump and arteries, veins, and capillaries are the tubes that carry the blood to and from the heart. Helps fight disease and clots after injury.

Respiratory System

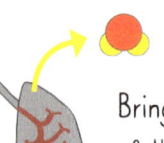

Brings oxygen into the body and gets rid of the carbon dioxide waste we make as we go through cellular respiration to break down food and release energy.

Taxonomy

? How do we keep all the types of life organized so humans across the globe can communicate regardless of spoken language?

QUICK QUESTION: What is taxonomy?

-> The science of naming and classifying organisms.

Dashing **DOMAIN**
King **KINGDOM**
Phil **PHYLUM**
Came **CLASS**
Over **ORDER**
For **FAMILY**
Good **GENUS**
Spaghetti **SPECIES**

We created levels of classification based on the relationships between organisms. Life forms with similar traits will be grouped together within each kingdom. We divide each kingdom into more and more categories, getting fewer and fewer organisms in each category until there is only one individual species based upon a unique characteristic. Like the individual leaves at the tips of branches.

Even though most organisms have a common name, taxonomy is in one language, Latin. We don't want to have to use all levels of the name, so we rely on just the last two; genus and species. This scientific name identifies a specific organism.

QUICK QUESTION: What is a species?

-> A group of organisms that can reproduce naturally with each other and create offspring who can do the same.

• Humans are "Homo sapiens" and a ladybug's scientific name is "Coccinella septempunctata".

WHOA!! That's a mouthful!

Here are some other ways we can classify animals to show how closely related they are to one another:

Phylogeny
A branching diagram, like a tree, that shows how organisms are related to one another over time.

Cladogram
A diagram that shows how organisms are related to one another based on physical differences.

Dichotomous Key
A tool designed to help identify different organisms based upon observable traits. You follow steps of two questions that will lead you to the organism.

Vertebrates With Jaws — 4 Legs — Dry Skin — Feathers

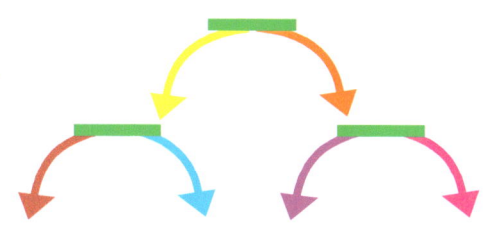

Classification Continued

Here is an example of a dichotomous key using different species of foxes:

Step 1

Is your fox small with very large ears?
GO TO STEP 2

Is your fox medium-sized with fluffy reddish fur?
GO TO STEP 3

Step 2

Does your fox have dark legs and a black tipped tail?
It is a Bat Eared Fox!

OR

Does your fox have light tan fur?
It is a Fennec Fox!

Step 3

Does your fox have dark legs and a white chin?
It is a Japanese Red Fox!

OR

Does your fox have a long snout and small ears?
It is a Tibetan Fox!

What do we use to determine how similar or different organisms are from each other?

1. Comparative Anatomy

Homologous structures are the same across species but are modified for different uses.

Vestigial structures have no use anymore.

Wisdom Teeth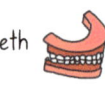

Goose Bumps

Appendix

Analogous structures developed to do the same job but are made differently.

 Embryos develop in similar patterns in animals.

2. Fossils show how species changed over time. This is the evolution of the horse!

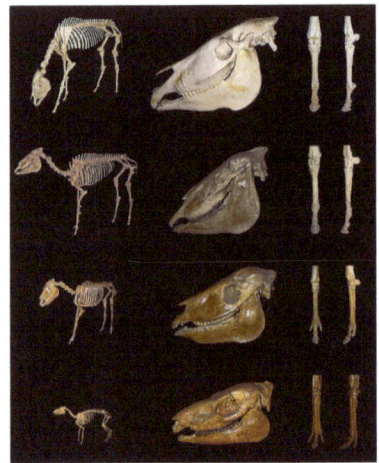

3. Biogeography studies how life moved across the planet.

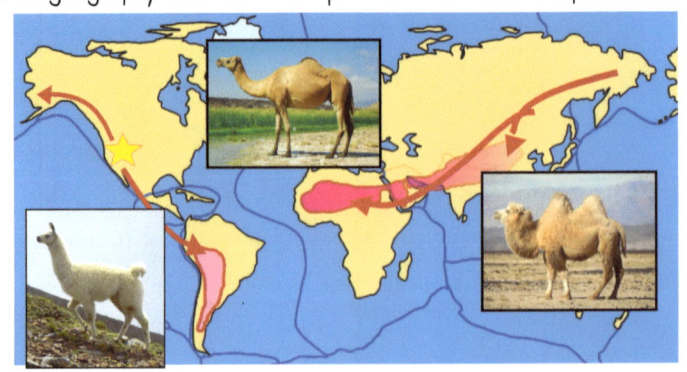

4. Biochemistry lines up the DNA code or the order of amino acids that make up proteins and compares them across species. Can you spot the differences?

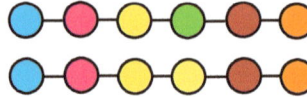

Data Dive

HUMAN	A	C	C	T	G	T	G	A	C	T
GORILLA	A	G	C	T	G	T	G	A	C	T
BIRD	A	G	C	A	G	T	C	A	C	G

The more DNA bases organisms have in common, the more closely they are related. Looking at the data, what organism is most closely related to a human?

Introducing Genetics

How and why does life look and behave the way it does? There are two arguments as to why, and ultimately, it is a combination of both!

< NATURE (Genetics) > VS < NURTURE (Environment) >

First let's discuss genetics, the nature behind a living organism.

QUICK QUESTION:
What is genetics?

-> The study of genes and how they are passed from parent to offspring.

QUICK QUESTION:
What is a gene?

-> A segment of DNA that contains the instructions for making a protein.

Humans have thousands of genes and the proteins produced make up everything you are! They not only make up your body parts, but some are specialized to move things, communicate messages, or even defend the body!

PROTEIN TYPES

Receptor (communication)

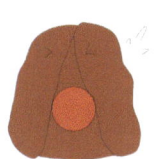
Contractile (squeeze)

Antibodies (defense)

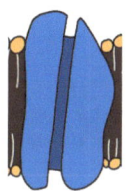
Transport (move)

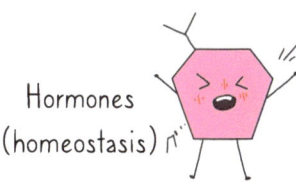
Hormones (homeostasis)

Storage (survival)

Enzymes (chemical reactions)

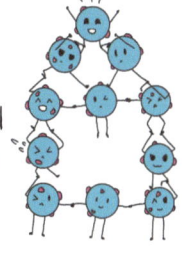
Structural (tissue)

Scientist Spotlight

Dr. Ambroise Wonkam was born in Cameroon and dedicated his research to understanding the genetics behind sickle-cell disease. Almost 75% of babies born with the disease live in Africa.

DNA can get changed in many ways producing new versions of proteins, which can be good or bad. The protein could no longer be made, still be made but not work at all, or it has a new role in the body. This is how adaptations are formed, slightly different versions of proteins where one type leads to better survival for the organism.

Mutations

There are many types of mutations, here is what can happen with DNA!

QUICK QUESTION:
What is a mutation?

-> A change in the DNA that is passed on to the offspring.

INSERTION
A nucleotide is added to the gene.

DELETION
A nucleotide is removed from the gene.

INVERSION
The DNA flips and inserts itself backwards.

SUBSTITUTION
A nucleotide is replaced by a different one.

Similar mutations can impact full chromosomes, which have thousands of genes changed at once!

Duplications are like insertions.

Deletions

Inversions

Sometimes chromosomes swap parts called translocation.

Humans have figured out how to use mutations to their benefit!

QUICK QUESTION:
What is genetic engineering?

-> A technique in which the DNA of an organism is altered.

QUICK QUESTION:
What is recombinant DNA?

-> DNA formed in a laboratory by combining DNA from two or more organisms.

Laboratory techniques using recombinant DNA have many uses!

Cloning gives us medicines like insulin.

GMOs (genetically modified organisms) can fight hunger issues with drought-resistant plants.

Gene therapy replaces bad genes with good genes; has worked for blindness!

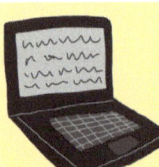

Ready to Research?!

What is DNA fingerprinting?
How do humans use it?

Search Suggestion: DNA fingerprinting, uses for DNA fingerprinting

DID YOU KNOW:
Bacteria have enzymes that cut DNA, which we use to cut and paste DNA together in the lab?!

Mendel VS Modern Genetics

Scientist Spotlight

Austrian monk Gregor Mendel discovered many principles of genetics, but little did he know that it got way more complicated! Since offspring are created from mother and father, he reasoned each offspring receives one copy of all the chromosomes from each parent, for a total of 2 copies. Here are some important terms related to what he found in his research:

DAD   MOM

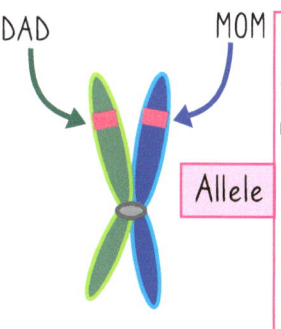

Allele

The version of the gene on a chromosome. Each chromosome is made from a copy from mom and a copy from dad. Therefore, you get two copies of every gene! Alleles code for the different forms of a trait (brown hair versus red hair for example).

Dominant Allele

 B ⬤ b

 B = brown hair

Mendel found that certain alleles dominate other alleles by showing up in the organism over the other form. We use a capital letter to represent this form of the trait.

Recessive Allele

b = red hair

The allele that is hidden by the dominant allele, or not seen in the organism. We use a lowercase letter to represent this form of the trait.

Genotype

A written form of the two copies of a gene. Knowing the alleles, whether they are dominant or recessive, allows us to predict what type of offspring that could be made from the parents. Here are the possible combinations:

BB   Bb   bb

MOM

	b	b
B	Bb	Bb
b	bb	bb

DAD

A punnett square is used to show possible children.

Phenotype

The physical description of the genes. All traits of an organism and their different forms that you can observe with your senses. (Brown hair versus red hair.)

With technology, modern study of heredity revealed some very interesting genetics!

Incomplete Dominance

Sometimes the two alleles blend together to make a new look. Neither one dominates the other!

HUMAN HAIR

curly   ✕   straight

wavy

Multiple Alleles

Sometimes there are more than just two alleles, there may be 3 or more. Fur color in rabbits have 4 alleles, just think about all those combinations!

HUMAN BLOOD TYPE

A and B alleles are equally dominant or co-dominant. They code for specific proteins that stick off your blood cells.

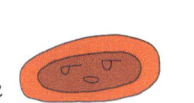 The O allele (no fancy protein) is recessive to both the A and B alleles.

AB blood type shows both A and B alleles.

Polygenic Traits

These are traits that are made from many genes combining together to produce the final look. This includes human height, skin tones, and eye color!

The Environment

Let's discuss the other side of the argument, or the theory that your environment impacts who you are when you grow up. How you are raised and where you are raised matters in many ways. Not only are you interacting with the physical environment like weather, but also the other forms of life within your community.

QUICK QUESTION:
What is an abiotic factor?

-> A non-living factor like climate and nutrients.

QUICK QUESTION:
What is a biotic factor?

-> A living organism like plants or bacteria.

Nutrients cycles follow the paths that certain important molecules travel throughout both the living and non-living parts of the environment. Some examples:

CARBON

HUMAN IMPACTS
1. burning fossil fuels
2. changing land use
3. ocean acidification

In the carbon cycle, the element carbon travels from air as part of carbon dioxide to the land as part of living organisms. Plants take in the carbon dioxide through photosynthesis and rearrange atoms to form the building blocks of life. Animals eat plants and use the carbon to build more of themselves. As living organisms die, the atoms of carbon get recycled into fossil fuels. As we burn fossil fuels, we are releasing carbon back into the air.

In the phosphorous cycle, the element phosphorous is released from rock formations through weathering. Plants take in the atoms from the soil and animals eat the plants to get the element. Decomposers release the atoms back into the soil which eventually is made into new rock.

PHOSPHOROUS

HUMAN IMPACTS
1. mining and erosion
2. sewage waste
3. detergents

NITROGEN

HUMAN IMPACTS
1. fertilizer run-off with algae blooms
2. air pollution
3. acid rain

In the nitrogen cycle, nitrogen is in the atmosphere and cannot be taken into plants or animals without a little help. Lightning causes nitrogen to pair up with oxygen in the air and these molecules can be taken in by bacteria. Different bacteria do separate chemical reactions that either make molecules that plants can absorb through their roots or put the nitrogen right back into the air.

Climate

Climate, or weather patterns over a long period of time, can greatly impact the number and type of organisms that live in a certain area. Climate data is usually a measurement of temperature and precipitation over time. Regions around the world have had a similar climate for a very long time. As a result, all living organisms have evolved adaptations for that area, usually due to the specific temperature range and amount of yearly precipitation.

QUICK QUESTION:
What is a biome?

-> A large area of land or water with specific types of animals, plants, and climate.

There are 5 major types of biomes, some are divided into more specific categories.

TUNDRA	
TEMPERATURE	-34°C to 12°C
PRECIPITATION	15 to 25cm

DESERT	
TEMPERATURE	-2°C to 26°C
PRECIPITATION	15 to 26cm

GRASSLAND	TEMPERATURE	PRECIPITATION
Savanna	20°C to 30°C	100 to 150cm
Steppe	-55°C to 45°C	25 to 50cm
Prairie	-40°C to 38°C	50 to 90cm
Pampas	1°C to 19°C	60 to 120cm

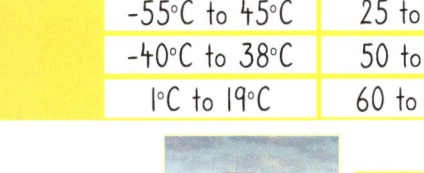

AQUATIC Saltwater (Marine)

AVERAGE OCEAN TEMPERATURE	4°C
AVERAGE OCEAN PRECIPITATION	250cm

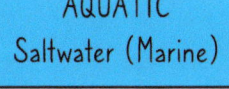

FOREST	TEMPERATURE	PRECIPITATION
Woodland (Chaparral)	0°C to 18°C	26 to 130cm
Temperate	-30°C to 30°C	50 to 150cm
Tropical	21°C to 30°C	150 to 1000cm
Boreal (Taiga)	-54°C to 21°C	30 to 85cm

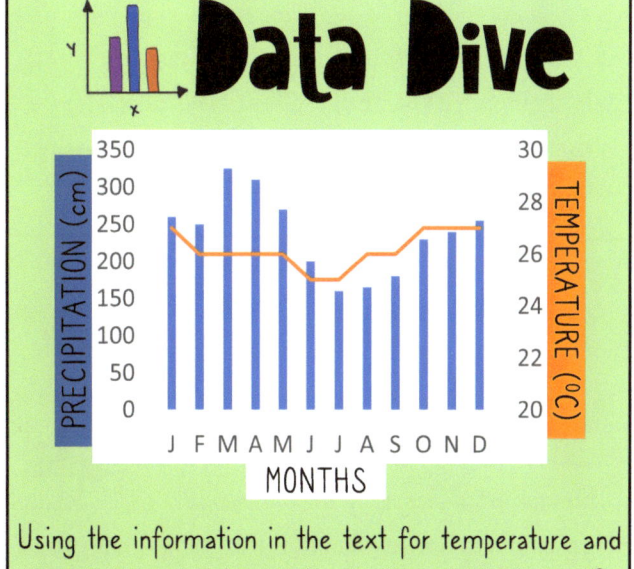

Data Dive

Using the information in the text for temperature and precipitation, what biome do you think this city is in?

29

Populations

QUICK QUESTION:
What is a population?

-> The members of a species that live in a particular area together.

You are part of the human population in your City!

When we look closer at populations in a biome, how did they get so successful and survive over time?

A man named Charles Darwin came up with an explanation after observing the different life forms across the Galapagos Islands.

Scientist Spotlight

Charles Darwin was an English scientist who came up with the theory of Natural Selection to explain how populations change over time, or what we call evolution.

QUICK QUESTION:
What is natural selection?

-> The process that a population goes through to adapt to a changing environment.

1. There are naturally occurring variations of traits (we now know this is the mutations in alleles of genes). Here are three colors of beetles:

2. Populations over reproduce, which leads to competition for survival.

3. Some of the offspring will have a variation that lets them out-compete the others. This adaptation allows them to survive longer and reproduce passing on their beneficial genes. "Survival of the fittest."

4. The beneficial adaptation is best in the environment, almost like it was chosen for the environment. We say this, adaptation is selected by the environment for survival.

This doesn't necessarily mean physically fit, it could be any behavior or structure that allows them to survive longer. Like camouflage!

Eventually the population will change over time where the successful trait shows up more and more in the population.

Population Limiting Factors

The environment has a limited amount of resources. Population size is controlled by a variety of factors. These factors limit population size to the environment's carrying capacity, or how many organisms the environment can support.

Density Dependent Factors
Depend on how many members of the population are living together.

Density Independent Factors
Do not depend on the population numbers, but the environment.

Disease

Weather or Climate

Natural Disaster

Predation

Competition

Human Disturbance

Factors in the environment go beyond just the population. Living organisms live in an ecosystem where communities of different species co-exist. Interactions between different populations can impact the success of a single population.

QUICK QUESTION:
What is an ecosystem?
-> A community of living organisms and their interactions with non-living factors.

QUICK QUESTION:
What is a community?
-> Many different populations living and interacting together in a specific area.

In communities, we can find very close interactions or relationships between species, called SYMBIOSIS.

MUTUALISM
Both of the species benefit from the relationship.

E. coli bacteria live in our gut and get nutrients from our food. In return they make vitamins, like vitamin K, which we use to clot blood!

COMMENSALISM
One species benefits, the other is NOT harmed.

Staphylococcus eats our dead skin cells for nutrients.

PARASITISM
One species benefits, the other is harmed!

Salmonella bacteria are parasitic bacteria that infect your digestive system.

Feeding Relationships

All organisms need energy, so the drive to get that energy has also shaped how communities form. We can summarize feeding relationships within a community in a couple of different ways:

FOOD CHAIN - a single set of feeding relationships

FOOD WEB - all possible feeding relationships

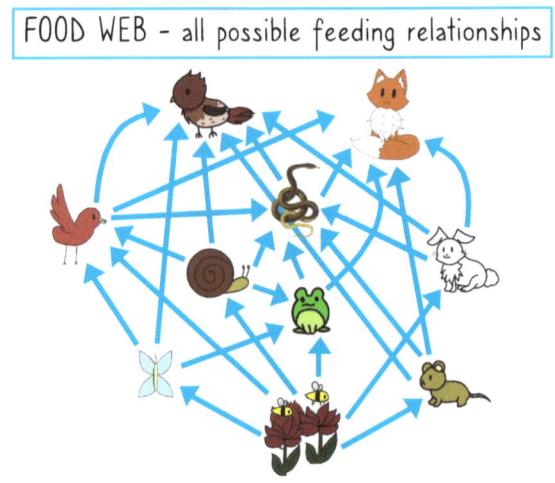

A trophic pyramid is used to show how food energy is passed through a food chain. The trophic levels represent the amount of energy available at each step, which gets less and less the further you get in the chain. In fact, only 10% of the energy is available to be used by the next organism!!

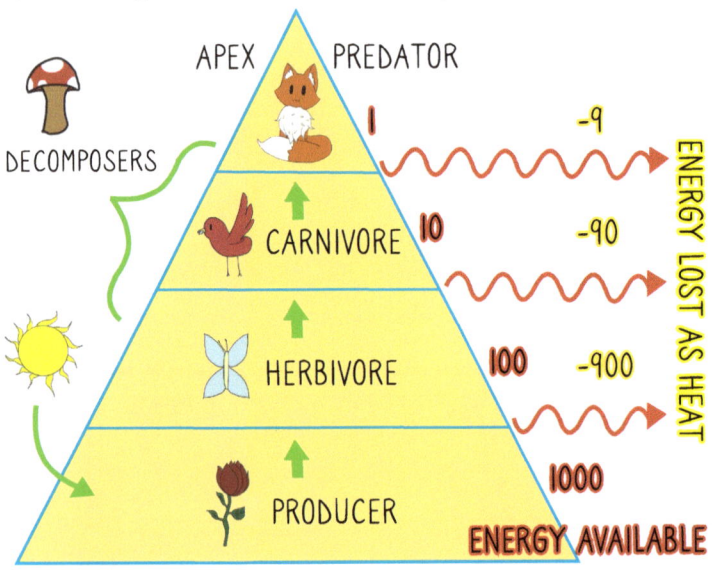

DECOMPOSERS

APEX PREDATOR

1   -9

CARNIVORE   10   -90

HERBIVORE   100   -900

PRODUCER   1000

ENERGY LOST AS HEAT

ENERGY AVAILABLE

1. Producers convert the sun's energy into sugar that is eaten by herbivores, or plant eaters.

2. Herbivores are eaten by carnivores, or meat eaters.

3. Some animals will eat both plants and other animals, they are called omnivores.

4. The top carnivore is called the apex predator as they are the animal that ends the food chain.

5. All living organisms die and their energy and nutrients are recycled back into the soil by decomposers, which is then taken up by plants.

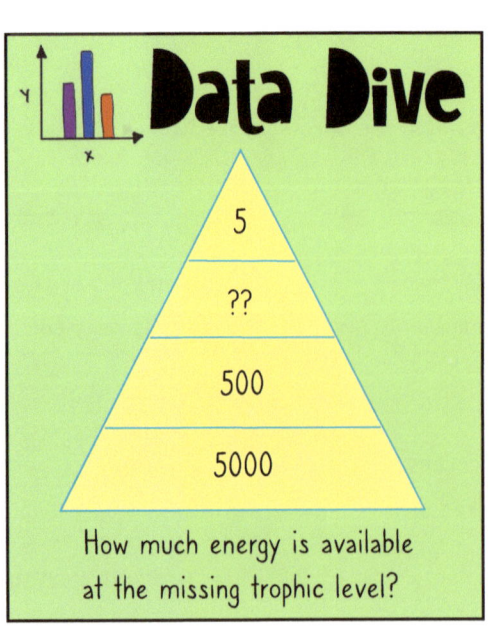

Data Dive

5

??

500

5000

How much energy is available at the missing trophic level?

15 eagles
150 frogs
1500 snails
15000 plants

As living organisms get energy, they also lose most of it to the environment as heat. This limits the amount of organisms that can be made at each tropic level.

CREATIVE CORNER

Create your own FOOD WEB from where you live. Share it to be featured on the HTH website!

submissions@howsthathuman.com

The Future

The biosphere is an amazing and unique place! We are searching for other planets, but so far, we live on a very special planet. It's just the right distance from our star, the sun. The right combination of molecules and minerals for life are in the Earth's crust and in the atmosphere. The atmosphere itself provides a protective barrier from harmful electromagnetic waves from space. Living organisms that have adapted to the pressures in the environment have survived and live in balance with each other.

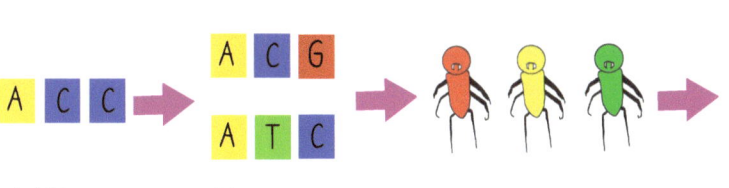

DNA code　　Mutations　　Variety of Traits　　Environmental Change

Life suited for the new environment survives, passes on genes, and takes over.

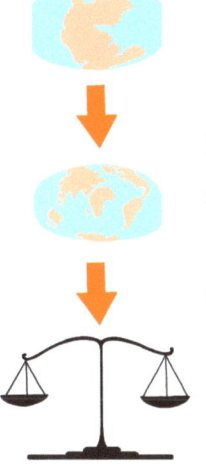

The earth has gone through many changes throughout its history. With natural disasters and climate change impacting the amount and type of life that survived each extinction event, life found a way to balance it out again.

Youth Spotlight

Eunbin Kang is a youth climate activist from South Korea. She co-founded the Youth Climate Emergency Action group in 2020. The group has already won court cases!

Both genetics and the environment contribute to what we see in the biosphere, but what are some threats?

Pollution makes the environment unsuitable for life.

Species overexploitation is the unsustainable hunting and gathering of life, like overfishing.

Invasive species outcompete natives for food and space and spread disease.

80% of deforestation is due to agricultural use.

Changes in land and sea use.

Life has adapted to current temperature and precipitation levels for years. Climate change alters the temperature, precipitation, or both and now life is vulnerable to extinction.

Importance of Biodiversity

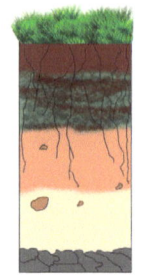

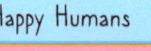

HEALTHY ENVIRONMENTS
Happy Earth

- clean air and water
- resilience to natural disasters
- soil formation
- ecosystem balance
- genetic diversity of life
- natural solutions to climate change

HEALTHY HUMANS
Happy Humans

- food security
- natural medicines
- mental health
- cultural/spiritual
- materials we need/use

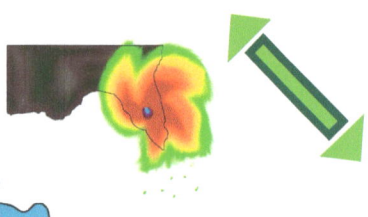

HEALTHY ECONOMIES
Successful Countries

- resistance to decline
- recreation
- tourism
- jobs

Hands-On From Home

Biomagnification Lab

Goal: To model how contaminants or dangerous chemicals can build up in a food chain.

Materials:
- shaker cup
- 9 small cups (minnows)
- 3 medium cups (sunfish)
- 1 large cup (osprey)
- 20 pieces of the same colored candy (producers)
- 10 pieces of a different colored candy
(producers with DDT chemical)

Procedures:
1. Place all the candies into the shaker cup and mix.
2. Pretend you are a minnow and you are eating 3 producers. Randomly place 3 candies in each of the 9 minnow cups.
3. Each sunfish, or the 3 medium cups, gets to eat 2 minnows. Randomly pour 2 minnow cups into each of the 3 sunfish cups.
4. The osprey, or 1 large cup, gets to eat 2 sunfish. Pour the candies of 2 of your 3 sunfish cups into the large osprey cup.
5. Did the osprey end up with any of the harmful DDT candies?

DDT is an insecticide that is washed into streams where it attaches to producers. Producers are then eaten by minnows who unfortunately take in the DDT themselves. Sunfish then eat the contaminated minnows, who then are eaten by the osprey. DDT prevents egg shells from forming properly, so the osprey populations decline due to fragile eggs.

What Can You Do?

As humans, you depend on the biosphere for your own survival!

-> The protection of natural resources, whether abiotic (non-living) or biotic (living).

-> Avoiding depleting natural resources to maintain an ecological balance. Meeting the needs of the present without compromising the needs of future generations.

1. Protect wild areas on land and marine habitats in the ocean.

As apex predators, sharks stabilize food webs. 71% of open ocean shark and ray populations have disappeared since 1970.

2. When designing cities make the space for nature.

Green architecture uses sustainable materials, energy consumption, and water usage.

3. Food and farming practices focus on a plant based diet.

Plant-based diets do not exclude meat or animal products, they just make plants the star of the dish.

4. Reduce climate change.

Bike more, drive less!

Use energy efficient light bulbs!

5. Protect water using non-toxic chemicals and conservation.

In 2023 the EPA (Environmental Protection Agency) updated water standards to include screenings for 6 PFAS, or forever chemicals, contaminating drinking water and harming humans.

6. Ecosystem management, which manages the use of an ecosystem so as to fulfill human needs without hurting the environment.

Wildlife crossings, like bridges or tunnels, are built to help humans travel through habitats.

7. Reduce, Reuse, Recycle!

8. Educate and volunteer.

When you learn more, you can help others learn more too! Organize a community clean-up!

"Some day the earth will weep, she will beg for her life, she will cry with tears of blood. You will make a choice, if you will help her or let her die, and when she dies, you too will die."
John Hollow Horn, Oglala Lakota, 1932

CREATIVE CORNER

What environmental threats do you face where you live? Research a topic, ways to improve it, and design a flyer or brochure to help educate your community! Share it to be featured on the HTH website!
submissions@howsthathuman.com

APPENDICES

Category Definitions and PBL Ideas

 QUICK QUESTION These questions are the main idea or key point of the section.

 Data Dive Practice math skills by analyzing data.

 Ready to Research?! Research prompts apply the information for further understanding of the topic.

 Scientist Spotlight Appreciate scientific contributions from around the world.

 Hands-On From Home Step-by-step experimental procedures, materials lists, and pictures.

 CREATIVE CORNER This category focuses on Project-Based Learning (PBL) projects.

 Check out below for additional ideas!

Brochure or Pamphlet	Poster	Start a Blog
Map	Presentation	Flipbook
Model	Diorama	Create a Cereal and Cereal Box
Treasure Hunt	Sculptures	Make a Time Capsule
Scavenger Hunt	Poem or Song or Rap	Advice Column
Menu	Timeline	Documentary
Advertisement	Write and perform a skit	Choose Your Own Adventure
Movie/Book/TV Show Review	TV or Radio Commercial	Top 10 List
PSA	Collage or Mobile	Museum Exhibit
News Report	Flowcharts or Diagrams	Paper Chain
Animation	Scrapbook or Photo Album	Album Cover
Stop Motion Animation	Instructional Video	Autobiography/Biography
Comic or Cartoon	Bookmarks	Newspaper or Magazine
Portfolio	Calendar	Invitation or Greeting Card
Awards	Casting Call	Sales Pitch
Banners	Episode of a Reality Show or Game Show	Book Club
Blueprints	Expert Panel Discussion	Puppet Show
App or Video Game Design	Children's Book	Debate
Board or Card Game	Fable or Myth	Mock Court Case
Design a Building or Garden	Help Wanted Poster or Ad	Create a Cheer
Cooking and Baking	Text Message Dialogue Box	Coat of Arms
Exercise	Series of Tweets	Flags
Plan A Lesson	Create a Social Media Page	Hieroglyphics
ID/Merit Badges	Murals	Tests/Worksheets
Illustrated Quotes	Pen-pals	Yearbook
Inventions	Stamps	Questionnaires

 Key Vocabulary

Abiotic Factor - A non-living factor like climate and nutrients.

Absolute Dating - This type of dating process uses the naturally occurring process of atoms breaking down over time.

Adaptation - A physical feature or behavior that allows an organism to survive better than others of its kind.

Biome - A large area of land or water with specific types of animals, plants, and climate.

Biotic Factor - A living organism like plants or bacteria.

Chromosome - DNA wound around proteins in tight coils.

Community - Many different populations living and interacting together in a specific area.

Conservation - The protection of natural resources, whether abiotic (non-living) or biotic (living).

Continental Drift - The theory that the continents have moved throughout earth's history.

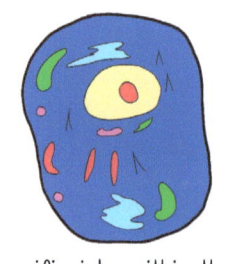

Ecosystem - A community of living organisms and their interactions with non-living factors.

Eon - a unit of time a half a BILLION years or more!

Era - A unit of time lasting several hundred million years.

Eukaryotic Cell - Cells with special parts called organelles. They have their own membranes and specific jobs within the cell.

Fossils - The preserved remains or traces of ancient organisms.

Gene - A segment of DNA that contains the instructions for making a protein.

Genetic Engineering - A technique in which the DNA of an organism is altered.

Genetics - The study of genes and how they are passed from parent to offspring.

Homeostasis - A state of balance for all body systems. This is needed for life to survive.

Invertebrate - An animal with no backbone.

Mass Extinction - a widespread decrease in the amount and type of living organisms.

Multicellular Organisms - Organisms made of many cells that take on specialized jobs.

Mutation - A change in the DNA that is passed on to the offspring.

Natural Selection - The process that a population goes through to adapt to a changing environment.

Osmosis - The movement of water molecules across a membrane.

Period - A unit of time lasting several tens of million years.

Population - The members of a species that live in a particular area together.

 # Key Vocabulary Continued

Prokaryotic Cell - A simple cell with no nucleus or specialized cell parts called organelles.

Recombinant DNA - DNA formed in a laboratory by combining DNA from two or more organisms.

Relative Dating - Using rock layers to date fossils, the oldest layers of rock are on the bottom.

Species - A group of organisms that can reproduce naturally with each other.

Sustainability - Avoiding depleting natural resources to maintain an ecological balance.

Taxonomy - The science of naming and classifying organisms.

Vertebrate - An animal with a backbone.

Honorable Mentions

This book has so many important vocabulary terms,
so here is a list of words to be familiar with from the text:

THE PAST

Biomolecule	Half-life
Nucleotide	Radioactive Decay
Carbohydrate	Organelle
Lipid	Diffusion
DNA	Active Transport
RNA	Hypertonic
Protein	Hypotonic
Amino Acids	Mitosis
Transcription	Photosynthesis
Translation	Cellular Respiration

THE PRESENT

Phylogeny	Insertion
Cladogram	Deletion
Dichotomous Key	Inversion
Homologous Structures	Substitution
Analogous Structures	Cloning
Vestigial Structures	GMOs
Climate	Gene Therapy
Biogeography	Allele
Nutrient Cycle	Dominant
Density Dependent Factor	Recessive
Density Independent Factor	Genotype
Competition	Phenotype
Predation	Polygenic Trait
Mutualism	Producer
Commensalism	Herbivore
Parasitism	Carnivore
Food Chain/Web	Omnivore
Carrying Capacity	Decomposer
Tropic Pyramid	Apex Predator

THE FUTURE

Biodiversity	Biomagnification
Pollution	Green Architecture
Species Overexploitation	Ecosystem Management
Deforestation	Climate Change
Invasive Species	Climate Activism

Scientific Method and Poster Design

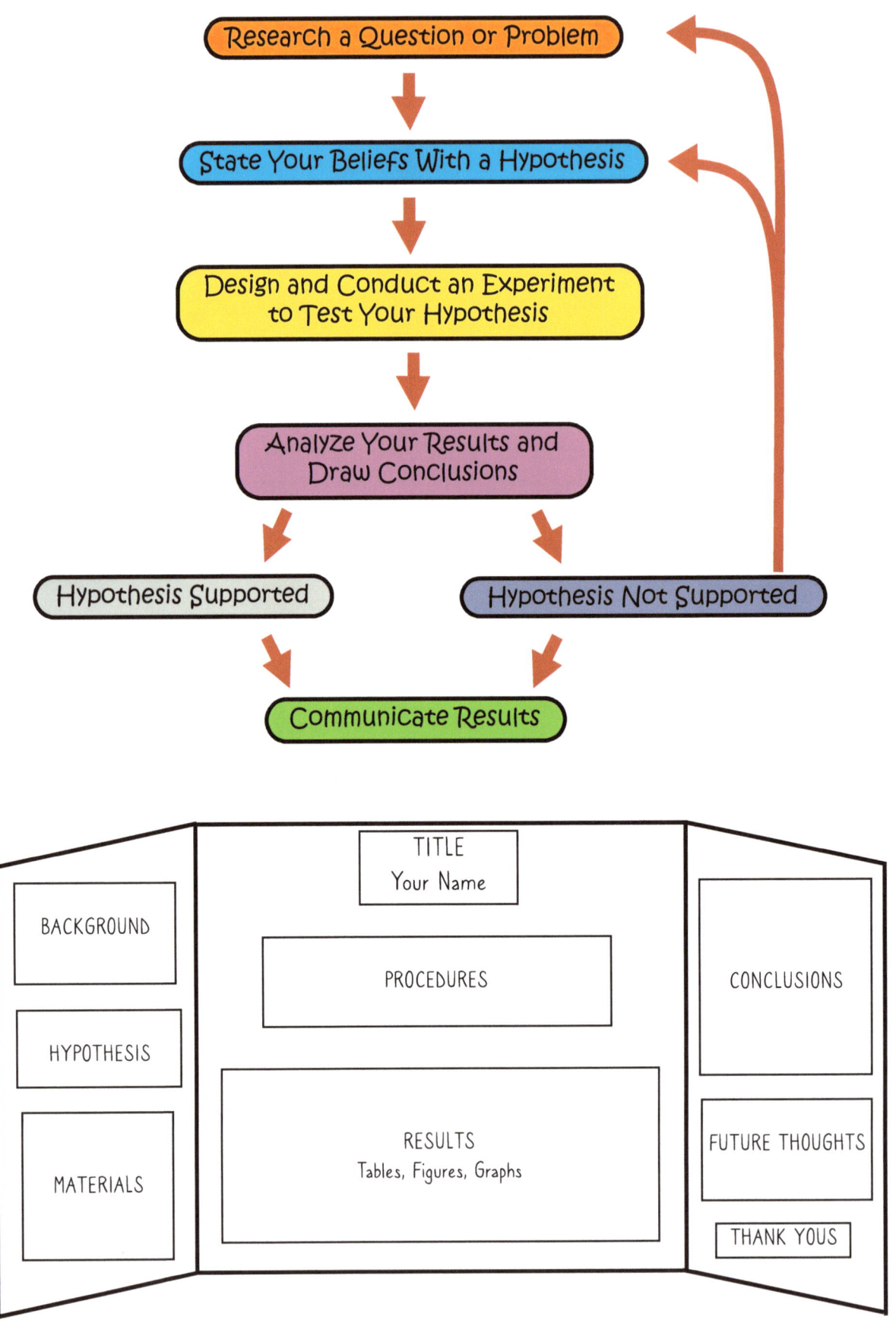

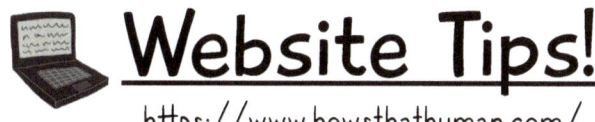

Website Tips!

https://www.howsthathuman.com/

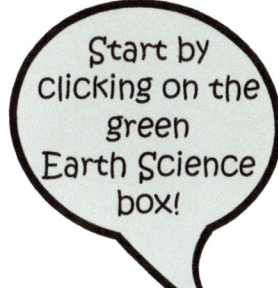

Start by clicking on the green Earth Science box!

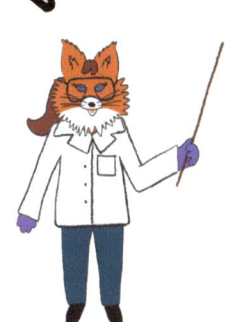

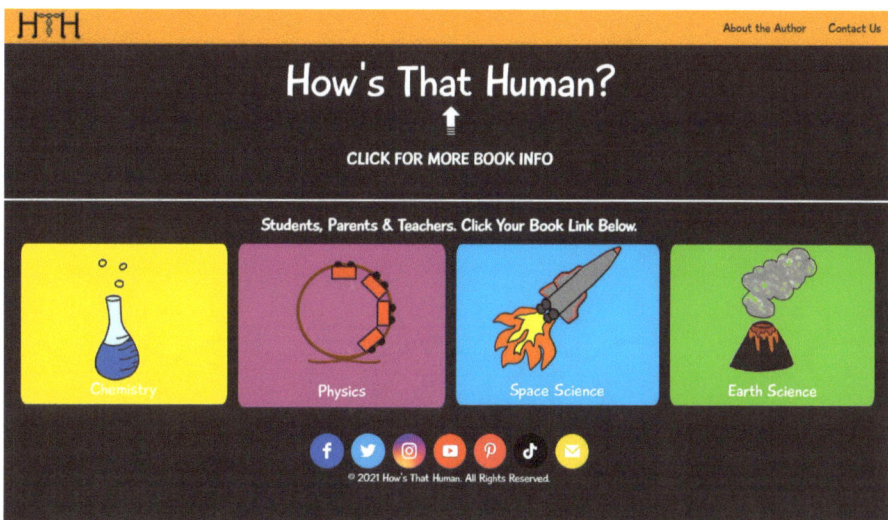

HTH
About the Author
Contact Us
How's That Human?
CLICK FOR MORE BOOK INFO
Students, Parents & Teachers. Click Your Book Link Below.
Chemistry
Physics
Space Science
Earth Science
© 2021 How's That Human. All Rights Reserved.

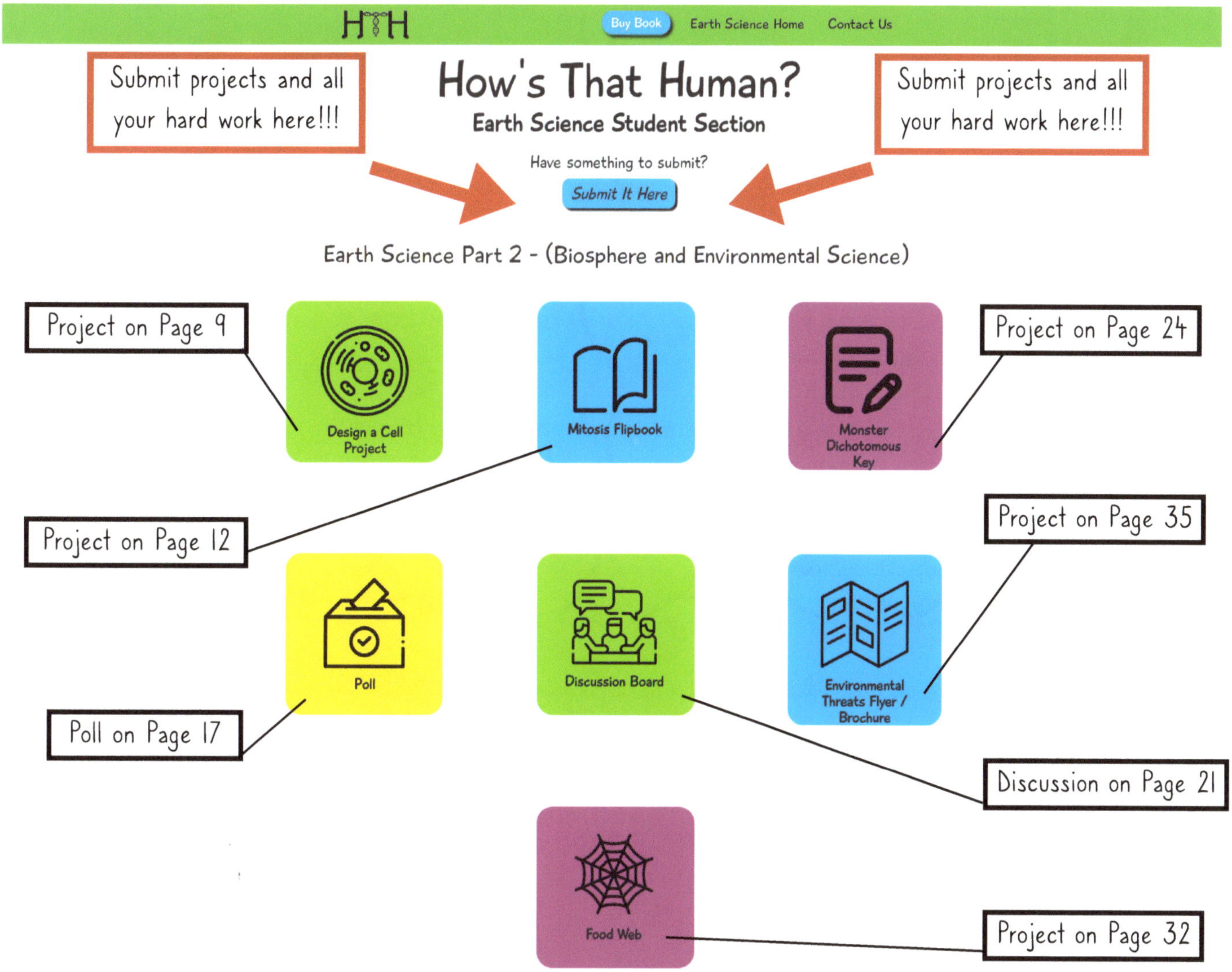
HTH
Buy Book
Earth Science Home
Contact Us
Submit projects and all your hard work here!!!
How's That Human?
Earth Science Student Section
Have something to submit?
Submit It Here
Submit projects and all your hard work here!!!
Earth Science Part 2 - (Biosphere and Environmental Science)
Project on Page 9
Design a Cell Project
Mitosis Flipbook
Monster Dichotomous Key
Project on Page 24
Project on Page 12
Project on Page 35
Poll
Discussion Board
Environmental Threats Flyer / Brochure
Poll on Page 17
Discussion on Page 21
Food Web
Project on Page 32

 # Author Spotlight

Author Rita Claire is a former molecular virologist and geneticist who worked on plant virus research involving cereal crops and human genetics projects. She misses research, but has thoroughly enjoyed teaching science K-12 and beyond. Ms.C, as she's affectionately called by her students, is passionate about changing the way we educate and has always dreamed of being an author. As a recent homeschool mom, she wanted to create a more relatable educational tool for other homeschool or distance learning families, as well as elementary school districts.
Welcome to How's That Human?

 # Artist Spotlights

Illustrator AM Conroy has a Ph.D. in biology from UCLA and has worked in marine biology, animal physiology, and biomechanics. Her love for animals inspired her to study the movements of amazing creatures such as swimming pufferfishes, running tigers, and hopping kangaroos. Her mom, an illustrator as well, encouraged her interest in drawing as a child, with her favorite subjects being snails, fishes, and unicorns. She enjoys illustrating for How's That Human? and hopes that her drawings help excite learners of all ages!

Illustrator Kas S. is a high school student who has been a student of Ms. C's for 2 years. Her interest in art started when she was young and she self-taught herself the subject. Kas loves to bring ideas and joy to others through her art and is always interested in trying out new art styles. Her love of art started when she was younger watching anime, cartoons, books, and manga. When Kas gets older she wants to get a degree in art or science. She's excited to be a part of How's That Human?

Acknowledgments

Pixabay	Pexels
NASA	Popsync
NOAA	Medium
FavPNG	H. Zell